# How to...
# SAVE THE
# PLANET

D0064254

# How to...

# SAVE THE PLANET

## By BARBARA TAYLOR

### Illustrated by
### Scoular Anderson

OXFORD
UNIVERSITY PRESS

*For Carissa, to help her save the planet*

# OXFORD
UNIVERSITY PRESS

Great Clarendon Street, Oxford OX2 6DP

Oxford University Press is a department of the University of Oxford.
It furthers the University's objective of excellence in research, scholarship,
and education by publishing worldwide in

Oxford  New York

Auckland  Bangkok  Buenos Aires  Cape Town  Chennai
Dar es Salaam  Delhi  Hong Kong  Istanbul  Karachi  Kolkata
Kuala Lumpur  Madrid  Melbourne  Mexico City  Mumbai  Nairobi
São Paulo  Shanghai  Taipei  Tokyo  Toronto

Oxford is a registered trade mark of Oxford University Press
in the UK and in certain other countries

Series devised by Hazel Richardson
Text copyright © Barbara Taylor 2001
Illustrations copyright © Oxford University Press 2001

The moral rights of the author and the artist have been asserted

Database right Oxford University Press (maker)

First published in 2001

All rights reserved. No part of this publication may be reproduced,
stored in a retrieval system, or transmitted, in any form or by any means,
without the prior permission in writing of Oxford University Press,
or as expressly permitted by law, or under terms agreed with the appropriate
reprographics rights organization. Enquiries concerning reproduction
outside the scope of the above should be sent to the Rights Department,
Oxford University Press, at the address above.

You must not circulate this book in any other binding or cover
and you must impose this same condition on any acquirer.

British Library Cataloguing in Publication Data available

ISBN 0-19-910740-8

3 5 7 9 10 8 6 4 2

Printed in Great Britain by
Cox & Wyman Ltd, Reading, Berkshire

Third party website addresses referred to in this
publication are provided by Oxford University Press
in good faith and for information only.
Oxford University Press disclaims any responsibility
for the material contained therein.

# Contents

# WHY THE EARTH NEEDS SAVING

When astronauts first set foot on the Moon, one of the most exciting things their photos showed was something most people thought they already knew about – Earth. When viewed from space it seems almost impossible to believe that such a tiny speck of rock could be home to such a fantastic variety of plants and animals – as well as over 6 billion people.

Viewed from space, it also becomes clear that the Earth is an island – it's out there on its own and we all depend on it for survival.

So, why does the Earth need saving? Well, to put it bluntly, because of us! At the moment the world population stands at over 6 billion (6,000 million). By the end of the 21st century, it may reach 8 or 10 billion. To satisfy the needs of these growing numbers of people, human beings are slowly but surely destroying the planet. We're using up the planet's natural resources, polluting the land, seas and skies, choking the air with fumes, clearing forests and other precious wildlife habitats to make way for more buildings. The way we live is damaging the very thing that gave us life in the first place – planet Earth.

I think we're in a no-win situation here!

This book tells you everything you need to know about the different problems facing the planet. After reading it, you'll know all about:

 how the Earth's atmosphere protects the planet

 holes in the ozone layer, and how they affect Earth and us

7

- why the Earth is getting warmer
- what pollutes the air and water
- what causes acid rain
- how we can save energy
- why cars and planes are bad news
- how waste can be recycled
- why it's important to save wild places such as rainforests

You'll also find special sections that tell you what you can do to help save the planet. Because everything in the world is interconnected, you'll find that a lot of the suggestions overlap. For instance, recycling paper cuts down on the piles of waste that pollute the Earth, but it also saves forests, because fewer trees have to be cut down to make new paper.

I'm helping to save the planet

# CHAPTER ONE

# HOLES IN THE SKY

Imagine it's a skin-blistering, eyeball-burning summer's day and you're sheltering from the heat under a sunshade. Then someone comes along and cuts holes in your shade-provider. Ouch! You're left sunburnt, sore and cross.

This is pretty much what's happening to the Earth. The Earth has its own built-in sunscreen, made of a layer of ozone, that stops too many of the Sun's ultraviolet rays from getting through the atmosphere.

9

These rays can cause all sorts of problems, from skin cancers to eye diseases (and not only in human animals). They are also bad for some crop plants, making them grow smaller leaves and suffer more from pests and diseases. Too many ultraviolet rays damage or kill off the microscopic sea creatures that feed fish, seabirds and the great whales. So holes in the ozone layer are pretty scary because they could eventually threaten all life on Earth.

## What is the ozone layer? .....................

Ozone is an invisible gas, which is actually a form of oxygen, the gas we need to breathe to stay alive. (Ozone has three oxygen atoms per molecule instead of the usual two.) It occurs high up in the atmosphere. The ozone layer is about 15–40 kilometres above the surface of the Earth. Without it, the Sun would frazzle us to a crisp.

Not quite crispy enough yet!

The reason the ozone layer is so fragile is that there is very little of it. The total amount of ozone between Earth and space is equivalent to a layer only 3 millimetres thick – although this is spread unevenly through the atmosphere, so it's a lot thicker over some areas. Most ozone is produced over the tropics, where the Sun's rays are strongest and most direct. This is because ozone can only be made with the help of the Sun's ultraviolet rays.

## What's making the holes?

A number of chemicals, which are released into the atmosphere by people, are causing holes in the ozone layer. Here's a list of the main culprits:

- In the past, chlorofluorcarbons (CFCs), which were used in aerosols, fridges and air-conditioners, destroyed a lot of ozone. They are now banned in Europe. Old machines containing CFCs have these chemicals removed before they are scrapped.

FREEZY FRIDGE
I'm just going to remove your CFCs – it won't hurt a bit!

Hydrofluorocarbons (HFCs) are replacing CFCs in some products. They only have between 2 and 10 per cent of the ozone-destroying power of CFCs, but they still destroy ozone.

Methyl bromide is a pesticide used by some farmers, especially those growing fruit and flowers.

When these chemicals reach the ozone layer, ultraviolet rays break them down, releasing chlorine or bromine. The chlorine or bromine joins up with one of the oxygen atoms from each ozone molecule, so destroying the ozone.

One molecule of chlorine can destroy thousands of molecules of ozone. As the ozone layer is weakened and thinned, it lets through more ultraviolet radiation, which speeds up the destruction of the ozone layer even further.

# Where are the holes? ........................

In the 1980s, scientists were shocked to discover that the amount of ozone over Antarctica had fallen by over two-thirds. The development of ozone holes over Antarctica happens for a few months each year, partly because of the long Antarctic winters. These produce frozen clouds high in the atmosphere, where ozone-destroying reactions occur.

As well as these temporary holes, the whole ozone layer round the globe seems to be thinning. In 2000, scientists found that almost two-thirds of the ozone layer over parts of northern Europe was lost for a few days.

The United Nations Environment Programme has estimated that a permanent 10 per cent reduction in the ozone around the globe will cause an extra 300,000 cases of skin cancer and 1.6 million cases of eye cataracts each year.

# Repairing the ozone layer ....................

Governments around the world have agreed to stop using CFCs and cut back on the use of other chemicals that contain the chlorine and bromine that damage the ozone layer. But there are still other problems.

Many of the alternatives to CFCs, such as HFCs, still contain some chlorine. And although there are some completely ozone-friendly alternatives, such as butane and propane, not everyone is using them. Many of the CFC alternatives have also been found to contribute to global warming and are already being phased out in some countries.

Even if all ozone-destroying substances were banned tomorrow, the CFCs already in the atmosphere will still hang around for 20 years or more, poking holes in the ozone layer.

## Rich and poor ....................................

In developing countries, such as China and India, more people are buying things like fridges and air conditioners. What's more, the types that they buy contain lots of CFCs. Governments in industrialized countries need to do as much as they can to help developing countries to use ozone-friendly alternatives.

I ♥ THE OZONE LAYER

# Go for green ....................................

There is a lot being done to tackle this issue already. But there are still some things you can do to help save the ozone layer.

 When your family or someone you know is buying a new fridge or freezer, make sure they ask in the shop for an ozone friendly one. Also get them to check that the CFCs in the old fridge are going to be drained out and recycled. Never just dump an old fridge. CFCs are only released when fridges are made and thrown away, not when they are being used.

 Check the labelling on stain removers, shoe cleaners, dyes, glues and paint-on correcting fluids and try to avoid ones with ozone-destroyers in them. (The ingredient to look out for is methylchloroform (1. 1. 1-trichlorethan).)

---

Don't forget to use sun hats, sunglasses and sun creams to protect your skin from the Sun's ultraviolet rays.

---

# CHAPTER TWO

# A WARMER WORLD

Have you noticed winters are getting warmer? Maybe you've been getting along with only three thermal vests where once you needed four?

Ever wondered why the planet seems to be hotting up? Well, it's no secret – it's called 'global warming'.

## What is global warming? .....................

Many scientists are convinced that the world is getting warmer because of human actions. We are releasing too much of some gases into the Earth's atmosphere. These gases trap some of the heat given off by the Earth, so that it can't escape into space. It's as if the Earth is trapped inside a huge greenhouse, because the gases trap heat in the same way as a greenhouse.

Greenhouse gases are produced by car exhausts, the burning of coal, oil and gas in power stations and factories, and rotting waste. They also come from – yes, you guessed it, CFC-containing fridges, air-conditioning units and foam plastics. (CFCs are 20,000 times better at trapping the Earth's heat than the most common greenhouse gas, carbon dioxide.) Most of the climate-changing pollution comes from industrialized countries, such as North America, Japan, Australia and those in Europe. Power stations and cars in these countries pump out vast amounts of carbon dioxide and other greenhouse gases.

## So what? .........................................

Well, at first it may sound like good news that the Earth is getting warmer. (Some scientists predict that the temperature could go up by as much as 4°C by the end of the 21st century.)

Those of you living in chilly countries probably think it would be nice if the weather warmed up a bit. But it's not as simple as that. A 4°C increase in temperature would make the world hotter than it has ever been during the whole of human history. This would create big, big problems all over the world.

❄ There will be bigger floods in coastal areas as the water in the oceans expands and glaciers and ice sheets melt even more than they already do.

- Sea levels could rise by 20 centimetres by 2030, and some islands in the Pacific Ocean will simply disappear under the waves.

- Areas once important for growing food, such as the American Midwest and the Ukraine, may become too dry for agriculture. Local food shortages, especially in Africa, could lead to higher food prices and more famine in Africa.

- The patterns of rainfall and winds could change, making the weather more unpredictable. This could also bring a lot more violent weather, such as typhoons, hurricanes, tornadoes and storms.

- Warm ocean currents like the Gulf Stream could change direction, making countries such as Britain colder instead of warmer.

- The changing climate conditions will make it difficult for some plants and animals to survive.

# Be an Earth Scientist:
## HELP TO REDUCE
## GREENHOUSE GASES

## WHAT YOU'LL NEED

- ☀ a young tree. (Choose a variety that grows naturally in your country.)
- ☀ a spade
- ☀ a stake
- ☀ string or something to tie the tree to the stake
- ☀ compost
- ☀ bark mulch

## WHAT TO DO

1   In autumn or spring, dig a deep hole in a place away from buildings where the tree will have plenty of space to grow. (Get permission from an adult first!) Some trees grow very tall and their roots spread out further than you think.

2   Put some compost in the bottom of the hole.

3   If the tree is over 1.5m tall, put a stake in the hole on the windward side of the tree.

4   Loosen the tree roots carefully and place the tree in the hole. (Make sure you get the tree the right way up!)

➤

5   Tie the tree to the stake and fill in the hole with soil.
    Stamp the soil down with your feet.
6   Soak the ground around the tree with about two buckets
    of water to encourage the roots to grow.
7   Sprinkle bark mulch onto the surface to hold the moisture
    and stop weeds growing.

## WHAT HAPPENS?

Trees (and other plants) soak up carbon dioxide and use it to
make their food and build new plant material. So, as it grows,
your tree will help to reduce the amount of one of the
greenhouse gases in the atmosphere and so cut down on
global warming!

# The story of the Earth's changing climate

Since the Earth formed some 4,600 million years ago, the climate has been constantly changing. At times it was a lot warmer than it is now. At other times, it was a lot colder, with much of the land covered in ice. These very cold times are called ice ages. The causes of changes to the Earth's climate include:

- continents drifting about the globe
- changes in the speed and direction of warm and cold water currents in the oceans
- changes in the amount of radiation given off by the Sun
- changes in the path of the Earth's orbit around the Sun, which changes the amount of sunlight falling on different parts of the Earth
- dust from space, meteorites hitting the Earth, or volcanoes on Earth blotting out the Sun and cooling down the climate. (This is one explanation for why the dinosaurs died out, together with more than half the species of plants and animals living 65 million years ago.)

## Climate change today .........................

One of the problems about climate change today is that it's difficult, even with the latest supercomputers, to predict how the climate of a whole planet will change in the future.

There are so many things to take into account, such as how fast the oceans will warm up and how much heat clouds will reflect and absorb. The only thing scientists are fairly sure about is that global warming is happening now and that we really should do something about it.

## What can be done? .........................

Global warming is such a big problem that it needs big solutions. Countries need to get together and agree what to do. Their governments can enforce changes, such as:

 cutting down on greenhouses gases and using sources of energy, such as wind, wave or sun power, which don't produce these gases

 removing carbon dioxide gas from power stations before it is released into the atmosphere

 burying carbon dioxide underground or in the oceans

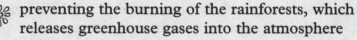 preventing the burning of the rainforests, which releases greenhouse gases into the atmosphere

 improving public transport so that people use cars less often

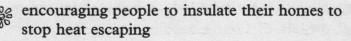 encouraging people to insulate their homes to stop heat escaping

# Go for green

Here are some things you can do to help slow down global warming:

 Use less electricity or gas so that power stations will pump less carbon dioxide into the atmosphere. Switch off lights and heating when rooms are not being used and use low-energy light bulbs.

Are you going out tonight?

 Use cars less. Take public transport if you can, or cycle, or walk. The last two things will keep you fit, too!

 Help to plant trees, which will soak up carbon dioxide.

# CHAPTER THREE

# AIR POLLUTION AND ACID RAIN

In Victorian times, large cities in Britain, such as London, Birmingham and Manchester, were covered by thick, yellow fog in the winter. These fogs were caused by smoke from coal fires in homes and factories, which built up in the still, cold air. The fogs were called 'pea-soupers' because they were thick and murky, like pea soup.

I wish they'd change the menu. I'm getting bored with pea soup every day.

Little was done to control the problem until December 1952, when a particularly bad fog settled over London. Traffic ground to a halt, London airport was closed and many people became seriously ill.

Still, it took another four years before the Clean Air Act of 1956 was passed by the government. This contained laws to control the amount of smoke produced by factories and homes. Many towns and cities were made into smoke-free zones.

## The air we breathe ............................

We've already seen how gases from cars, factories and power stations can cause global warming. There is also the added problem of the air pollution that they cause. A lot of this is from invisible gases such as carbon monoxide, sulphur dioxide, nitrogen oxides and ozone. You're probably breathing in some of these right now!

Traffic fumes and other forms of air pollution can make illnesses like asthma much worse. However, asthma has been around since long before cars were invented. The Ancient Egyptians used to treat it with crocodile poo!

## Be an Earth Scientist: 
## CHECK LOCAL 
## POLLUTION LEVELS

One way to tell how polluted the air is in your area is to carry out a lichen survey. Lichens are strange living things that usually look like flat green or orange omelettes stuck to stones or tree bark. Sometimes, they look like beards. Lichens are made of a fungus and an alga living together.

Lichens are very sensitive to air pollution. If you can't find any, your air is badly polluted.

Crusty          Leafy          Beardy

Crusty lichens will stand some pollution.
Leafy lichens will only stand a small bit of pollution.
And beard-like lichens will only grow in clean air.

## Killer rain

Air pollution isn't just a problem in the air. It also mixes with water vapour in the atmosphere to make acid rain. All rain is naturally slightly acidic, but gases from power stations and vehicle exhausts make it much more so. These strong acids eat away at the stone of buildings and statues and make them crumble away. Acid rain also damages trees and weakens them by changing the chemistry of the soil around their roots. This means the trees are less able to cope with disease, insect attack and other problems. Trees with needle-like leaves are the worst affected.

About 50 per cent of the trees in the UK, Germany and the Netherlands have damage which might have been caused by acid rain. There is dispute over whether damage is caused directly (on the needles) or indirectly (by affecting resistance to disease, and so on).

Rivers and lakes are also affected by acid rain. In Sweden some 4,000 lakes have virtually no fish at all, while another 20,000 have been badly affected, although some are slowly recovering.

# Be an Earth Scientist:
# CHECK ACID LEVELS IN
# YOUR RAIN

## WHAT YOU'LL NEED
* a red cabbage
* a saucepan
* a sieve
* a jug
* three jars
* vinegar
* bicarbonate of soda
* rainwater

## WHAT TO DO
1 Chop up a few red cabbage leaves and put them in a saucepan with half a litre of tap water.
2 Ask an adult to help you boil them gently for ten minutes.
3 Let the mixture cool, then pour it through the sieve into the jug to get rid of the bits of cabbage.
4 Pour a small amount of the cabbage water into the three jars.
5 Add a few drops of vinegar to one jar, a teaspoon of bicarbonate of soda to the second jar and some rainwater to the third jar.

## WHAT HAPPENS?
Acids, such as vinegar, will turn the cabbage water pink. If your rainwater is very acidic, it will turn the cabbage water a strong pink colour. Bicarbonate is an alkali (the opposite of an acid) and will turn the cabbage water green. (This is a great experiment, but don't try drinking the results!)

I said fruit tea, not vegetable tea!

# Air action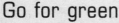

Governments can take action to reduce the amount of air pollution in a number of ways. They can require factories and power stations to use more efficient boilers and furnaces that produce less pollution. Also, fuels such as coal and oil can be processed so that they produce less pollution when they burn. For instance, fitting Flue Gas Desulphurisation (FGD) to power stations reduces the amount of sulphur dioxide ($SO_2$) they pump into the atmosphere.

Adding lime to soils, lakes and rivers is one way of making them less acidic. However, this is very expensive and scientists don't really understand how this changes the chemistry of the water or soil.

Catalytic converters (CATs) in cars reduce the amount of pollution in their fumes. And cars that use fuel more efficiently create less air pollution in the first place.

# Go for green

Here are some things that you can do to lessen air pollution and the effects of acid rain:

 Use less energy, especially electricity for lighting, heating and running machines. This will mean less pollution from power stations.

Cycle to school, or arrange a 'walking bus', where you meet up with other children and all walk to school safely together.

 Keep a regular check on the lichens in your area, to be sure that air pollution isn't getting worse.

 Buy second-hand toys or do some swaps with friends for a change. This would mean factories made fewer new things, and less pollution would be pumped into the air.

# CHAPTER FOUR

# WATER POLLUTION

Where would we be without water? Well, nowhere actually. Without water there would be no life on this planet. Water covers almost three-quarters of the Earth's surface.

You'd think that with so much water about, it would be hard to do much damage to the oceans and seas, wouldn't you? Yet all too often, the oceans are used as a giant dumping ground for oil, sewage sludge, toxic chemicals and radioactive waste. Nowadays, we are pumping in so much pollution that some seas, such as the Baltic, the North Sea and the Mediterranean can't dilute the poisons quickly enough. Many beaches have become too polluted for safe swimming.

## A threat to life .....................................

In 1989, pollution from rivers such as the Po in Italy created huge mounds of slimy green algae in the Adriatic Sea between Italy and Yugoslavia. Sewage pollution encourages algae to grow in vast numbers, using up all the oxygen and killing fish and other water life.

The fresh water in rivers and lakes and underground water is also threatened by the poisons leaking from waste dumps, chemicals from industry and pollution from farms, such as pesticides and fertilizers that are washed into the water.

# Oops!
# Oil spills at sea

You know how much mess it makes if you accidentally drop a glass of juice? Well, oil spillages are much worse! Since 1975, enough crude oil has been spilt to fill almost 1200 Olympic-sized swimming pools.

They don't mean that literally, stupid!

Here's a list of some of the biggest spillages.

- 1978: the *Amoco Cadiz* spilled 233,000 tonnes of oil off the coast of France.
- 1989: 38,000 tonnes of oil were spilt into Prince William Sound, Alaska from the *Exxon Valdez*. Between 260,000 and 580,000 seabirds and hundreds of seals, bears, deer, mink and river otters were killed. A total of 1,700 km of shoreline was eventually affected by the oil.
- 1993: the *Braer* spilt over 84,000 tonnes of oil off the Scottish coastline.
- 1996: the *Sea Empress* lost about 73,000 tonnes of oil in Pembrokeshire, South Wales.
- 1999: the *Erika* sank in a gale off the coast of Brittany, France, releasing at least 11,000 tonnes of oil and killing more than 58,000 seabirds.

But oil tanker accidents are only responsible for a small percentage of the oil polluting the world's oceans. Much more of the oil released by ships comes from tankers cleaning out their tanks. Other ways that oil gets into the sea include oil and gas exploration and drilling at sea, oil refineries on land, and people carelessly getting rid of engine oil.

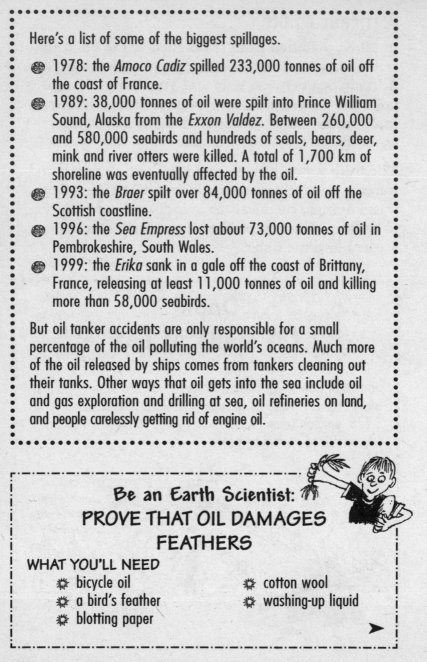

**Be an Earth Scientist:**
## PROVE THAT OIL DAMAGES FEATHERS

**WHAT YOU'LL NEED**

- bicycle oil
- a bird's feather
- blotting paper
- cotton wool
- washing-up liquid

➤

## WHAT TO DO

1   Lay the feather on the blotting paper and put a few drops of water on to the feather. Look carefully at the shape of the water drops.
2   Dry the feather and put a few drops of bicycle oil onto the feather. Spread the oil over the feather with the cotton wool and then drop some water on top. What shape are the water drops now?
3   Put some washing-up liquid and water into a bowl and wash the oiled feather. How easy is it to get the oil off?

## WHAT HAPPENS?

The clean feather is waterproof so water forms round drops and rolls off the surface. On the oily feather, the water soaks in. Oily feathers stick together and do not keep out the cold and wet. When birds try to clean their oily feathers, they swallow the poisonous oil. Even with soapy water, it's not that easy to clean feathers. Some oiled seabirds can be cleaned up after an oil spill, but many are covered in too much oil to be saved.

# Pulling the plug ...............................

Have you ever wondered what happens to your bath water when you pull out the plug? Or where all the poo goes when you've flushed the loo?

It usually goes to a sewage treatment works to be broken down and cleaned up before the water is put back into rivers or the sea.

Special bacteria can be used to eat up the sewage and destroy harmful dirt and germs. But often the waste does not get enough treatment and pollutes rivers or the sea. Some countries do not have sewage treatment works at all.

Sewage poisons wildlife. It uses up oxygen as it breaks down, leaving less for the wildlife in the water. It also contains millions of bacteria and tiny creatures that cause infections and diseases.

39

# Be an Earth Scientist:
## MAKE YOUR OWN
## WATER FILTER

**WHAT YOU'LL NEED**
- ✹ a large plastic bottle
- ✹ a large plastic cup
- ✹ scissors
- ✹ cotton wool
- ✹ gravel
- ✹ sand
- ✹ blotting paper

**WHAT TO DO**
1. Make a mixture of muddy water.
2. Cut the bottom off a large plastic bottle and wedge a piece of cotton wool in the neck of the bottle.
3. Turn the bottle upside down and fix it firmly in the plastic cup.
4. On top of the cotton wool, add a layer of stones and gravel and a layer of fine sand.
5. Put a piece of blotting paper on top of the sand.
6. Pour your muddy water carefully through the filter.

➤

**WHAT HAPPENS?**
Some of the dirt sticks to the blotting paper, sand, gravel and cotton wool so the water will be cleaner when it comes out of the bottom of the filter. But don't drink the water! It will still contain dirt and germs that could make you ill.

## Dirty water, clean water .......................

The amount of nasty stuff that gets tipped into our rivers, lakes and seas by big companies, farmers and individuals needs to be reduced or cleaned up so that it doesn't pollute the water.

Better sewage treatment could make a difference, too. But it's not always easy to find out who has caused the pollution or make people obey water pollution laws. And improving sewage treatment works costs money.

# Go for green ....................................

Here are some things you can do to help save the world's water:

 Save water by doing things like taking a shower instead of a bath and turning off the tap while you are brushing your teeth.

 Don't use more bubble bath or shampoo than you need to. They pollute the water and make it harder to clean.

 Put a water container in the garden to catch rainwater for the plants, instead of watering them with tap water.

 Don't pour oil or paint down the drain – suggest that an adult takes them to a local tip to be recycled.

 Encourage adults not to use pesticides in the garden, so that they won't get washed into the water underground.

 Don't dump litter in the sea, on beaches or in rivers.

 Report any water pollution you see to your local council or to the water company.

42

# CHAPTER FIVE

# ENERGY

We need energy for everything we do – for running, jumping, walking, talking and sleeping. Even the worst couch potatoes need some energy to reach for the remote control or lift another chocolate bar to their mouths!

*Why do they waste all that energy chasing a small round object they never even eat?*

We get the energy we need from the food we eat. You burn the food inside your body to release energy.

But from the earliest days of human history, people have used other sources of energy, such as wood to burn on fires, as well.

43

Nowadays, people in industrialized countries (such as Europe and North America) use vast amounts of extra energy to heat and light their homes and run machines such as washing machines, vacuum cleaners, televisions, computers and cars. North America has only 4 per cent of the world's population, but uses 25 per cent of the world's energy. India, on the other hand, has 17 per cent of the world's population but uses only 2 per cent of the world's energy.

## Where does energy come from? ...........

Our main source of energy is the Sun. Every half hour, the Earth receives more energy from the Sun than is released by all the coal, oil and gas burned in the world for a whole year.

Plants use the Sun's energy to make food. Animals get their energy by eating plants, or by eating other animals. Over millions of years, the dead remains of plants and animals have been turned into coal, oil and gas. These are called 'fossil fuels', because they were formed from the fossilized (preserved) remains of plants and animals.

# DON'T MISS THIS
## ONCE-IN-A-LIFETIME CHANCE TO

# TRAVEL BACK IN TIME

TO THE COAL FORESTS OF 300 MILLION YEARS AGO.
PREPARE TO BE DIVE-BOMBED BY DRAGONFLIES
AS BIG AS SEAGULLS AS YOU WALK OVER THE DEAD LEAVES
AND WOOD THAT, OVER MILLIONS OF YEARS, WILL BE
PRESSED TOGETHER TO FORM COAL.
THEN DIVE DOWN TO THE OCEAN DEPTHS IN OUR SPECIAL

# MINI-SUBMARINE

TO WATCH OIL AND GAS FORMING FROM
THE DEAD BODIES OF PLANTS AND ANIMALS
RAINING DOWN ON THE SEABED.

## HURRY!
# FREE
SOUVENIR LUMP OF COAL TO ALL THOSE
WHO BOOK BEFORE 1ST APRIL.

So when we burn fossil fuels in power stations or in our homes, we release the Sun's energy stored in animals and plants that lived long ago. Unfortunately, we also add to the problems of global warming and acid rain.

## Energy on the run ..............................

The trouble with fossil fuels is that once they are used up, that's it. We can't get them back. And we're getting through them at frightening speed. If we continue to use fossil fuels at the same rate as we do now, this is roughly how long they will last:

 coal 200–300 years

oil 40–60 years

natural gas 65 years.

*They ran out of coal, so I bought this wind turbine instead.*

Coal provides about a third of the world's energy today and is one of the main fuels used in power stations to make electricity. Natural gas provides about one fifth of the world's energy.

Oil supplies about half of all the energy we use. If all the barrels of oil produced in a day were laid end to end, the line would stretch twice around the equator.

## Nuclear power

Some power stations use nuclear power (energy from splitting atoms) to make electricity. Nuclear power stations don't produce greenhouse gases, or cause acid rain. But they do have a couple of major drawbacks.

# Accident at Chernobyl

### Chernobyl, Ukraine, 1986

The accident at Chernobyl nuclear power station in 1986 was far worse than any oil spill. An explosion in the nuclear reactor killed 32 people and injured hundreds of others. And thousands of people in the area are still suffering from the after-effects of the disaster.

Clouds of radiation from the explosion spread over the whole of Europe and Scandinavia. Huge numbers of plants and animals were contaminated. Even reindeer in northern Scandinavia and sheep in the UK were killed, as a result of eating lichen that had been poisoned by the radioactivity.

➤

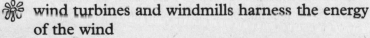

The risk of dangerous accidents at nuclear power stations is one reason that many people are against producing electricity in this way. Another reason is that nuclear power stations produce very dangerous radioactive waste. This gives off radiation so deadly that it must be kept out of harm's way for hundreds, possibly thousands of years. At present there is no safe way of doing this.

## Renewable energy ...............................

There are alternative ways of making electricity, from energy sources that are cleaner, safer and won't run out:

 solar power uses the power of the Sun

wind turbines and windmills harness the energy of the wind

water power uses the power of falling water

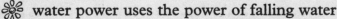

48

- wave power traps the energy of the waves
- tidal power harnesses the energy of the tide as it comes in and goes out
- heat from underground rocks (geothermal power) can be used to make electricity
- biogas is a fuel source that can be made from animal dung and human sewage.

These alternative energy sources are better for the environment in many ways, but they still have their problems. Wind turbines, windmills and solar power stations take up a lot of land; setting up tidal power stations can destroy wildlife habitats; and to harness geothermal power means drilling into rock, destroying land and causing pollution. Also, energy from alternative sources can be expensive and unreliable.

## Be an Earth Scientist:
## MAKE A SOLAR PANEL

### WHAT YOU'LL NEED
* black plastic bin bag
* a shallow cardboard box
* sticky tape
* scissors
* clear plastic tubing, about 8 mm diameter
* modelling clay
* thin wire
* wire cutters

### WHAT TO DO
1   Line the box with the black plastic, holding it in place with sticky tape.
2   Cut a piece of plastic tubing, about three times the length of the box.
3   Fix the tubing inside the box in an S shape. Use small lengths of wire, poked through the sides of the box, to hold it in place.
4   Seal one end of the tube with a blob of modelling clay and fill the tube with water. Make sure the unsealed end of the tube is pointing upwards when you add the water.   ➤

5    Leave your solar panel in the Sun for a few hours and then
     tip the water out into a bowl.

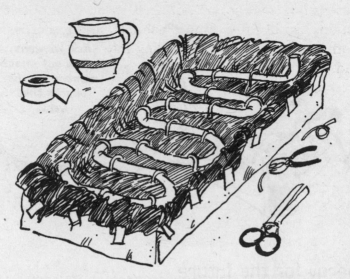

**WHAT HAPPENS?**
The energy from the Sun should heat up the water in the tube
just like a real solar panel. The black plastic absorbs the Sun's
heat and helps the water to warm up.

## The wind-up radio ...............................

Radios are usually powered by batteries or plugged
into mains electricity. It takes energy to make batteries
and, when they are thrown away, the metals leak out
and cause pollution. An inventor called Trevor Baylis
has invented a radio that does not need electricity or
batteries. To make his radio work, you just wind it up
for about 20 seconds and it plays for about an hour.

As well as saving energy, the wind-up radio can be used by people who do not have an electricity supply and can't afford batteries.

# Energy for the future ........................

To stop our use of energy causing so many problems, such as global warming and acid rain, governments could put more money into renewable forms of energy, such as solar, water or wind power.

But the best way to reduce our energy problems would be to use less energy in the first place. Governments could encourage people to insulate their homes, fit a jacket around the hot water tank and have their windows double-glazed, so that heat energy stays inside the home and doesn't escape.

## Go for green ......................................

Here are some things you can do to save energy:

 If you're cold, put a jumper on, don't turn the heating up. And if you're hot, turn the heating down, rather than opening windows while the heating is still on.

 Persuade your parents to buy low-energy light bulbs. Some energy-efficient light bulbs use 80 per cent less energy than ordinary ones and last eight times longer. So energy-efficient bulbs help the environment and save money!

 Switch the TV off at the wall, not by remote control. A colour TV left on standby can still use a quarter of the energy it uses when it's on.

 Dry washing outdoors instead of using a tumble-drier.

When boiling a kettle, only put in the amount of water you actually need, and use it straight away.

 Use your hands instead of energy-guzzling machines. Wash up instead of using a dishwasher, and brush your teeth with an ordinary (not an electric) toothbrush.

# CHAPTER SIX

# TRANSPORT

Today, there are hundreds of different types of transport that move people and goods around, from bicycles and roller blades to cars, buses, trucks, aeroplanes – and even space shuttles. If all the cars in the world were parked end to end, they would stretch around the equator over 36 times.

# Travelling in time

Until about 250 years ago, the main forms of transport were feet (walking!), sailing ships and animals (horses, donkeys, camels). But in the 18th, 19th and 20th centuries, a whole load of inventions changed the face of transport forever.

- Early 1700s Horses pulled railway coaches along tracks.
- 1783 The first hot-air balloon took off.
- Early 1800s Steamboats and trains were developed.
- 1863 The first underground train system in the world was built in London, England.
- 1880s First cars went on sale to the public in Europe and the first electric trains were tried out.
- 1900–1910 First aeroplanes developed.
- 1904 First mass-produced car, the Ford Model T, invented in the United States.

*No, that's not what I meant by a model T!*

- 🌑 1935 First motorway built, in Germany.
- 🌑 1961 First space flight.
- 🌑 1990s Very large cargo ships transport goods over the oceans. Each ship uses enough power to light half a million light bulbs.

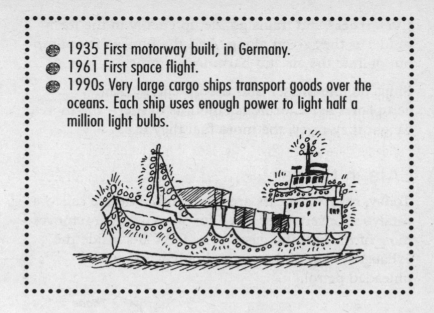

## The trouble with transport ...................

Most forms of transport have a bad effect on the environment.

*Something bothering you?*

Just making cars, trains, ships and planes in the first place uses up energy and materials. Then there's all the land that has to be cleared to make way for roads and railways.

Cars, trucks and trains guzzle up energy in the form of fuel as they zoom along. And the fumes they pump out pollute the air and harm living things.

Ships pollute the sea and use up energy. And aeroplanes use enormous amounts of fuel – the more weight they carry, the more fuel they use.

## CATs in the car ...............................................

Today, many new cars are fitted with a device called a catalytic converter (or CAT for short), which removes most of the harmful gases. The CAT fits inside the exhaust system of the car and only works with unleaded petrol.

Sounds good doesn't it? But of course, there's a snag. CATs do not remove carbon dioxide, so the car fumes still contribute to global warming. Also, CATs are made of rare metals and digging these metals out of the ground damages the environment. Yet another problem is that the CAT won't begin to work unless the engine has warmed up, so on a short trip, it probably doesn't work at all.

# Future cars ........................................

Scientists are looking for new kinds of car fuel that will not pollute the environment. One day cars may be able to run on solar power – using the Sun's energy to power car batteries. At the moment, however, these cars are just experiments, but who can say what the future will bring?

Another source of fuel might be the hydrogen taken from water. There is an awful lot of water on the Earth, from which the hydrogen can be made. And when hydrogen burns, the exhaust gases are steam, not polluting gases.

Cars powered by electric batteries are another possibility. But batteries aren't powerful enough for long journeys, and the power to recharge them has to come from somewhere. If the power comes from a polluting power station, then the cars will still be damaging the environment. Batteries are also toxic when they are thrown away.

Cars can run on fuels such as ethanol or methanol, made from plants such as sugar cane and oilseed rape.

Fuels made this way would be sustainable, because more plants could be grown to replace the ones used up. But large areas of the Earth would have to be taken up growing these plants, which might not be good news for the environment.

## Top transport tips ..............................

Governments can do a lot to encourage people not to use their cars so much. If public transport were cheaper, more reliable and comfortable, people would use it more. Sixty cars, each carrying one person, use about 16 times more energy than one bus carrying 60 people.

Governments could also:

 Encourage individuals and businesses to use trains more, especially for long-distance travel and carrying goods.

Railways are cheaper to build and maintain than roads, and trains use fuel more efficiently than cars.

 Make cycling safer by setting up cycle lanes in every town and city.

 Stop companies building out-of-town shopping centres, because most people have to drive to them.

Build new houses near city centres and places where there are jobs, so that people don't have to travel so far to get to work.

Make driving cars more expensive by putting up the price of petrol and car parks. The money made from these extra charges could be spent on public transport.

 Only allow cars in city centres at certain times of day.

# Go for green .....................................

Here are some things you can do to cut down on the problems transport causes:

 If you're planning a holiday or other long trip, get your family to go by train. It's better for the environment and can be more fun, too.

 Get your school or council to provide safe places to leave bikes, so that more people can cycle to school.

 If your family or friend plans to buy a new car, encourage them to get one with a catalytic converter (CAT).

 Suggest that your parents and friends use unleaded or low sulphur ('city') petrol. Lead and sulphur are both poisonous to people and pollute the air.

 Buy local! Use local shops and buy food that is grown locally to cut down on the journeys needed to transport people and goods to and from the shops.

 If the car you are in is stuck in a traffic jam for a long time, ask the driver to switch off the engine.

# CHAPTER SEVEN

# THAT'S RUBBISH!

Every month, each of us throws away our own body weight in rubbish. That's not just left-over chips, old coke cans and mouldy fruit, but newspapers, plastic packaging, old toothpaste tubes, clothes and shoes that are too small or out of date, batteries that have run out, broken electrical goods, old toys ... the list is endless.

Somebody who likes playing around with numbers has worked out that every European will leave behind a mountain of waste that is roughly 1,000 times their body weight by the time they die. For Americans, the figure is some 4,000 times their body weight. For someone from Madagascar, the amount is only 100 times their body weight.

The solution to the waste mountain is not for us all to move to Madagascar. We need to stop throwing things away and find ways to re-use them, mend them or give them to someone who can use them. What would be better still is if we all stop buying so many things in the first place!

## What's wrong with new things? ...........

Making new things uses up resources. Materials such as metal and oil (most plastics are made from oil) have to be dug out of the ground. This leaves nasty gashes in the landscape and destroys and pollutes wild places. What's more, these kinds of material cannot be replaced once they have been used up.

Transporting the raw materials to factories also uses energy. The factories that make the goods use up energy too, usually from fossil fuels, which add to the problems of global warming or climate change.

Then the goods are wrapped in a lot of unnecessary packaging (more waste), and transported to the shops (more pollution), to be used and thrown away on the rubbish mountain.

Sure it's nice to have new things, but we could make much better use of the things we have. Even small changes can make a big difference.

You might like to think about this interesting little fact: if each of the UK's 10 million office workers used just one less staple a day (by re-using a paper clip) an incredible 120 tonnes of steel would be saved each year.

## Dead bodies and poo ........................

The natural rubbish produced by plants and animals (dead leaves, rotting wood, animal bodies, animal poo – that sort of thing) is never wasted. Lots of living things, especially bacteria and fungi, find waste really yummy. Dead leaves make tasty snacks for worms.

As waste gobblers like these feed, they break down the waste so that it can be used to build new living things and help them to stay alive.

But natural rubbish buried on rubbish tips can produce a gas called methane. This gas can make rubbish tips explode. It also causes global warming.

WHUMP

In the Netherlands, 90 per cent of the natural rubbish from people's homes is recycled; in Denmark 55 per cent is recycled, and in Austria the figure is 50 per cent. Other countries could follow these good examples.

---

**Be an Earth Scientist:**
## RECYCLE YOUR RUBBISH:
## MAKE A COMPOST HEAP

If you have a garden, you can turn your old food waste and grass clippings into free fertilizer that helps plants grow.

➤

---

## WHAT YOU'LL NEED

- ❋ a bucket
- ❋ fruit and vegetable peelings
- ❋ lawn clippings and leaves
- ❋ rabbit, guinea pig or hamster poo
- ❋ weeds

## WHAT TO DO

1   Collect your waste in the bucket and pile it up in the corner of the garden. (You could ask an adult to help you make a compost bin from wood and chicken wire to stop the rubbish spreading over the garden.)

2   Keep the compost heap moist (but not wet) and ensure it is loose enough for air to circulate, so that living things can survive there.

3   You can cover the heap with a piece of old carpet or wood to keep it warmer so that it will turn into fertilizer more quickly.

## WHAT HAPPENS?

Bacteria, fungi and creatures such as worms, millipedes, slugs and earwigs feed on the heap and make it rot down into rich compost. All this feeding makes the heap heat up – within a week it may even be steaming!

➤

Compost takes up to six months to form, so don't expect things to happen overnight. Then you can dig it into the flower beds or the vegetable patch to make the soil rich. This is heaps better than buying bags of peat, which may have been dug up from peat bogs (important habitats for rare plants and animals).

## Worm power

If you haven't got a garden, or enough room to make a compost heap, you can still make compost using a worm bin and some special compost worms, called brandling worms. You can get the worms from someone else's compost heap, fishing shops or in kits. Use the compost for your pot plants or give it away.

## Wrap it up

About a third of the rubbish we throw out is packaging. Packaging is important – it keeps food fresh, makes freezing food possible, stops things getting squashed or broken and gives us information.

Your breakfast cereal would be pretty difficult to carry back from the supermarket without its box. And baked beans without the tins just don't bear thinking about.

But a lot of the packaging we end up bringing home isn't really necessary. It just makes products look bigger and better than they really are. Easter eggs are a good example of this.

HAPPY EASTER

One of the most common sorts of packaging is the plastic bag we use to carry things home from the shops. The handles cut into your hands like knives, and the bags often split open half-way home.

You're much better off with a rucksack, basket, box or canvas bag.

## Once is not enough ..............................

Getting rid of rubbish by dumping it in holes in the ground or burning it in incinerators causes huge problems and harms the environment. The obvious answer is to recycle more of our rubbish. This saves using up raw materials, cuts down on the energy we use and reduces pollution.

You'd be surprised at what can be recycled. Hopefully, you already recycle your old newspapers, cans and bottles. You can do all sorts of things with them, if you use your imagination.

But there are also places or organizations that recycle greetings cards (including Christmas cards), spectacles, video games, batteries, books, clothes, shoes, computer software, toners for photocopiers, print cartridges for computer printers and many other things.

## Paper

You probably know that paper is made from trees. But did you also know that every year, each one of us uses up two trees' worth of paper and cardboard? At that rate, it won't take long to use up all the world's forests, together with the wildlife and peoples that live there. And making new paper may involve using chlorine bleach, which pollutes rivers.

# Be an Earth Scientist:
# MAKE RECYCLED PAPER

## WHAT YOU'LL NEED

* scrap paper
* bowl
* liquidizer or potato masher
* four pieces of thin wood
* hammer, nails and drawing pins
* fine curtain netting, wire gauze or the backing used for tapestry weaving
* blotting paper
* rolling pin
* an iron

## WHAT TO DO

1   Tear up the scrap paper and leave it to soak in a bowl of hot water until it's really soft and mushy. (You could add coloured paints, tinsel, wool, leaves or even pencil sharpenings to make your recycled paper look more interesting.)

2   While the paper is soaking, make a sieve to strain the water from the mushy mixture. Nail the four pieces of wood together to make a frame and cover this with the net, mesh or gauze. Fix the frame cover in place with drawing pins or nails.

➤

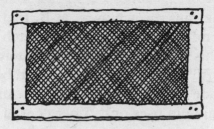

3   Put some blotting paper on top of several layers of newspaper.

4   Mash or liquidize the soaked scrap paper.

5   Dip the sieve carefully into the mashed-up mixture and hold it flat above the bowl to let the water drain through the sieve.

6   Turn the sieve quickly upside down and tip the layer of mushy paper onto the blotting paper.

7   Put another piece of blotting paper on top of the mushy paper and use a rolling pin to squeeze out as much water as possible.

8   Ask an adult to help you iron the blotting paper until the recycled paper is nearly dry.

9   Peel off the top piece of blotting paper and leave your recycled paper in a warm place to finish drying.

## WHAT HAPPENS?

The fibres in the old paper spread out in the water. When you drain off the water, the wet fibres knit together to make a criss-cross network of fibres. This then dries into sheets of recycled paper without any glue to hold the fibres together.

## Cans

If all the drink cans thrown away in Britain every year were placed end to end, they would stretch beyond the Moon. This terrible waste is completely unnecessary. All food and drink cans can be melted down and made into new cans. Making cans from recycled aluminium uses only 5 per cent of the energy needed to make cans from raw aluminium. It is especially important to recycle aluminium cans because they take much more energy to make than steel cans do.

## Glass

Glass is made mainly from sand and limestone – and there's hardly a shortage of them in the world. But digging sand and limestone out of the ground leaves great holes in the landscape and causes pollution. More importantly, making glass uses enormous amounts of energy. When a tonne of recycled glass is used to make new glass the equivalent of 135 litres of oil is saved.

This is partly because recycled glass melts at lower temperatures than sand and limestone. But recycling also saves the energy needed to dig up the sand and limestone and deliver them to the glass factory.

If we just throw our old bottles and jars away, they clog up rubbish dumps and last forever, broken and buried in the soil. Recycling or re-using old glass bottles and jars changes all that.

## Plastics

Plastics are a problem. Most plastics are made almost entirely from oil. Every year, as much oil is used up as it takes nature 1 million years to create. Plastics factories spew out polluting waste into rivers and sewers. One plastic, PVC, is a major source of dioxin pollution when it is burnt. Dioxin is a chemical that is harmful to health, even in tiny amounts.

When most plastics are thrown out, they don't rot away (although some plastic bags are now being produced that are designed to break up slowly). This means it's especially important to find ways of recycling and re-using plastics.

Plastics are difficult to recycle, partly because there are so many different types. Mixed plastic waste can be recycled to make certain products, such as fence posts and traffic cones, but plastics really need to be sorted and recycled separately to make high-quality products. It would be easier to recycle plastics if the industry set up more sorting facilities for used plastics. It would also help if each product was made of one type of plastic, and if plastic products were clearly labelled so that people could easily sort them.

Dairies in the USA, Canada and Sweden use plastic milk bottles that can be reused up to 100 times each. Perhaps in the future more products could be packed in reusable bottles.

# Go for green

Here are some things you can do to cut down on the amount of stuff you throw away:

 Save paper. When you are writing or drawing, try to use both sides of the sheet whenever you can, and re-use old envelopes. Make your own greetings cards, instead of buying new ones. Use recycled writing and loo paper. Recycle your waste paper.

 Save cans. Take flasks or reusable bottles on picnics or hikes. Recycle old cans. Wash them out to stop them smelling and squash the sides together so that they take up less space.

Save glass. Take your old glass to a bottle bank, or better still reuse it for something else.

 Save plastic. Buy glass containers instead of plastic ones, when you can. Use plastic bottles that can be refilled at the shop you bought them in. Reuse or recycle your old plastic containers, for example as flower pots, for mixing paint or for storing things. Avoid carrier bags, but if you have to get one, try to reuse it in some way.

# CHAPTER EIGHT

# SAVE WILDLIFE AND WILD PLACES

In this chapter we're going to look at the dangers facing the wildlife and wild places on this planet. First, try this true or false quiz to see how much you know and to find out some amazing facts.

## True or false? .....................................

1    An area of rainforest the size of a football pitch is cut down every minute.

2    The Arctic is the only place on Earth that is not contaminated by poisonous chemicals.

3    Islands are especially at risk from habitat destruction. (Habitats are places where plants and animals live.)

4    The world's most endangered cat is the tiger.

79

5   Nearly half of all crocodilians (crocodiles, alligators, caimans and gharials) are endangered.

6   Dodos were killed off by global warming.

7   Rhinos are rare because people kill them for their tusks.

8   Orang-utans will be extinct in the wild in five to ten years if rainforest destruction continues at its present rate.

9   Up to 80 per cent of Britain was once covered in wild woodland; now only 1 or 2 per cent is covered in trees.

10  Sharks are more likely to be attacked by people than the other way around.

Answers to True or False QUIZ

1  False. An area of rainforest the size of six football pitches is cut down every minute

2  False. There is, alas, no place on Earth that is free from poisonous chemicals, not even the bottom of the oceans.

3  True. Islands are home to small numbers of unique wildlife that can be wiped out very quickly.

4 False. The Iberian lynx is the world's rarest cat. There are only a few hundred left and it's teetering on the brink of extinction. Tigers are not doing very well either though. There are probably only 2,000–3,000 of them left.

5 True. They are threatened by loss of habitat, illegal hunting, competition with people catching the fish they eat, and egg collection.

6 False. Dodos became extinct because people hunted them, and rats and dogs ate their eggs.

7 False. Rhinos don't have tusks!

But rhinos are killed for their horns, which are made into dagger handles and used in traditional Chinese medicine.

8 True. There are now only about 5,500 orang-utans left on Sumatra and 8,000–12,000 on Borneo, largely because the rainforests where they live are being cut down and forest fires also threaten them.

9 True. British woodlands have been cleared by people over the last 4,000 years and now Britain is one of the least wooded countries in Europe.

10 True. Only about 24 people are killed by sharks every year, but millions of sharks are hunted for their fins (for shark's fin soup), meat, skin and cartilage, while many more are killed accidentally in fishing nets.

## Going, going, gone ............................

Only a tiny fraction of all the life that has ever existed on the Earth has been preserved in some way – in amber, stuck in natural tar, deep-frozen in the soil or trapped between layers of rock.

These remnants from the past tell us that the pattern of life is for new living things to develop, last for a few million years and then die out, or become extinct. They are probably killed off by competition from other living things or changes in the Earth's climate.

Over the last 400 years, living things have been dying out at a much faster rate than they usually do. One estimate is that the natural rate of extinction is one species every 100 years, while the speeded-up rate (caused by people) is one species every 30 minutes! At this rate, half of the planet's inhabitants could disappear during the next century.

# Why are species disappearing? .............

The main reason for the great extinctions of recent years is the great increase in numbers of people – people who need somewhere to live, grow their crops and keep their farm animals, people who build roads and take resources such as metals, trees, oil and coal from the Earth.

Other threats to wildlife include:

 people killing animals for their fur, skins or other body parts (such as horns and tusks)

 pollution of the air and water caused by things such as acid rain, pesticides and increased ultraviolet radiation getting through the ozone layer

 global warming or climate change

 over-collection of animals for the pet trade or plants for gardens

- new species (such as cats and rats) being introduced to an area and destroying the wildlife already there

- diseases passed on from people to animals (gorillas have been badly affected by diseases such as measles, and colds caught from people).

## Does it matter? ...................................

At this point, some of you may be wondering why we should bother to save all these wild places and the creatures that live in them. People have got to live and eat haven't they? Well, yes, but... it just so happens that we are not living in a vast empty space, but on a living, breathing planet. We are connected to all the living things around us through the food we eat, the air we breathe, the water we drink and the soil we grow things in. Protecting the variety of life on Earth (the 'in-word' for this is biodiversity) helps to keep the whole planet healthy and helps us to survive at the same time.

On a selfish note, plants and animals that we have not even discovered yet may provide us with life-saving medicines and food in the future.

On an unselfish note, do we have the right to get rid of other living things, just because they don't suit our lifestyle? Don't they have just as much right to live as we do?

# Be an Earth Scientist:
## BUILD A POND

**WHAT YOU'LL NEED**
- ☀ permission to dig the pond
- ☀ a spade
- ☀ some large stones or bricks
- ☀ plastic sheeting
- ☀ old cloth or carpet
- ☀ old pond water or tap water
- ☀ soil
- ☀ pond weeds

**WHAT TO DO**
1. Choose a spot away from overhanging trees, so that the pond won't get clogged up with leaves in autumn.
2. Ask an adult to help you dig a hole about 2 m deep and 2 m wide. The sides should rise in a series of shallow steps.

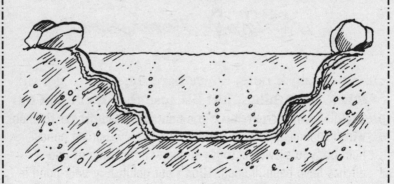

3. Clean out any sticks or stones from the bottom of the hole, put some old cloth or carpet in the bottom and line the hole with strong plastic.

➤

4   Hold the plastic in place with bricks or stones.
5   Cover the bottom of the pond with a layer of soil about
    10 cm deep and fill with old pond water or tap water
    to about 10 cm from the top (the level will go up when
    it rains).
6   Leave the water for a week and then add some pond
    plants on the ledges, together with some stones for
    animals to shelter under. Leave an area of sloping soil on
    at least one edge so that animals can get in and out of the
    pond.

## WHAT HAPPENS?

Your pond should develop into a mini-ecosystem where you can
watch changes in the web of life day by day. Garden ponds
help frogs to survive because so many natural ponds and
ditches have been drained. (Don't put goldfish in your pond if
you want frogs and newts. The fish will eat their eggs and
tadpoles.)

Note: If you cannot make a proper pond, an old sink or plastic
bowl sunk in the soil or a tub of water on the patio will do
instead.

# Be an Earth Scientist:
# MAKE A BIRD TABLE

## WHAT YOU'LL NEED
- ☀ a square of wood 30 cm x 40 cm
- ☀ four thin strips of wood to go along the sides
- ☀ glue
- ☀ paintbrush
- ☀ wood preservative
- ☀ string or nylon thread

## WHAT TO DO
1. Glue the four strips of wood along the edges of the square, leaving a gap in each corner for the rain to drain out.
2. Paint the wood with a wood preservative that is not harmful to wildlife.
3. Ask an adult to help you put two screw eyes at the base of each short side.
4. Thread string or nylon thread through the holes.
5. Hang your bird table from a branch or windowsill. (Keep it away from bushes where cats could leap out and attack the birds.)
6. Put food such as bread, nuts and seeds on the table.

## WHAT HAPPENS?
In cold weather, the food on your bird table could save lives. Once you start feeding the birds, do so regularly because they will come to rely on your food. In warm weather, there is plenty of natural food about so that there is less need to feed the birds.

# Conservation measures ........................

Governments and companies control how land is developed and can play a major part in preventing habitat destruction. Industrialized countries have destroyed much of their own wild places, yet often seem to be telling other countries not to do the same thing. Developing countries need help, not lectures. Many people think that the debts such countries owe to industrialized countries should be cancelled in exchange for preserving their wild places. Cutting back on pollution in all countries will also help to preserve habitats.

Other things governments could do are:

 set up more national parks and nature reserves

 give people control of their own local environment (such as forests) to encourage them to protect their surroundings

 ban more of the trade in rare species

 catch poachers that try to capture rare species

 breed rare species in zoos or botanic gardens and re-introduce them back into the wild preserve seeds for the future by freezing them in seed banks

 plant trees that naturally grow in an area to replace those that are cut down

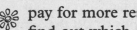 pay for more research to identify new species and find out which ones are endangered.

## Go for green.......................................

Here are some things that you can do to save wildlife and wild places:

 Don't buy ivory, fur, coral or bone jewellery or any other goods made from rare species.

 Buy dolphin-friendly tuna. Many dolphins die when they are trapped in some tuna fishing nets.

 Visit nature reserves and raise money for conservation organizations.

 Avoid disturbing plants and animals in the countryside and don't pick or dig up wild plants.

 Make bird tables and ponds to encourage wildlife in your garden.

 Don't use chemical pesticides in the garden, and use only organic fertilizers.

 Persuade your family not to buy furniture made from rainforest trees, such as mahogany or teak, unless it comes from a certified sustainable source.

# CHAPTER NINE

# HOW GREEN ARE YOU?

Now for the big test: let's see how green you are with this quiz. No cheating now: be honest!

1   Does your family use a tumble-drier instead of drying clothes outside?
2   Does your family intend to recycle the CFCs in their old fridge when they buy a new one?
3   Have you ever planted a tree?
4   If it is too hot in a room, do you:
    a turn down the heating
    b open a window?
5   Do you get to school by:
    a walking, cycling or public transport
    b car
    c sharing a car?
6   Does your family car run on unleaded or low sulphur ('city') petrol?
7   Do you turn off lights and heating in empty rooms?
8   Does your family use low-energy light bulbs?

9    Do you leave the TV on standby?

No, I never turn the TV off at all.

10   Is your house insulated and double glazed to stop heat energy escaping?

11   Do you usually wash in:
     a  the shower
     b  the bath?

12   Do you leave the tap running when you clean your teeth?

13   Do you recycle:
         a  glass bottles     b  plastic bottles
         c  newspapers        d  cans?

14   Do you:
         a  use recycled paper
         b  reuse envelopes?

15   Do you ever drop litter?

16   Do you use:
         a  new shopping bags every time
         b  your old bags that you take shopping with you?

17   Do you buy goods with as little packaging as possible?

18 Does your family buy organic products when it can?

19 Do you have, or have you helped to create:
    a a pond
    b a bird table
    c a compost heap
    d a wildflower patch?

20 Do you support or raise money for 'green' organizations (those that work to protect the planet)?

## Scoring

1 yes 0, no 1.
2, 3 yes 1, no 0.
4 a 1, b 0.
5 a 2, b 0, c 1.
6, 7, 8 yes 1, no 0.
9 yes 0, no 1.
10 yes 1, no 0.
11 a 1, b 0.

12 yes 0, no 1.
13 a 1, b 1, c 1, d 1.
14 a 1, b 1.
15 yes 0, no 1.
16 a 0, b 1.
17 yes 1, no 0.
18 yes 1, no 0.
19 a 1, b 1, c 1, d 1.
20 yes 1, no 0.

**Green rating for quiz**
**1–9** Oh dear. You're about as green as an overripe strawberry, aren't you? Well, at least you've made a start by reading this book. Now that you know more about the problems facing planet Earth and how to solve them, maybe it's time to do your bit to help!

**10–18** Not bad. You're already doing a lot to save the planet, but you could do more. Maybe you just needed a few ideas to get you going on a greener lifestyle. Don't give up!

**19–28** Wow, you're really doing your bit to save the planet – keep up the good work, and spread the word among your friends and family!

## Can I really make a difference?..............

Saving the planet is an enormous task. The problems are huge and complicated and a lot of the time even the scientists don't fully understand what is happening. So it's hard to predict what will happen to the Earth in the future. The amount of time people have lived on the Earth is like the blink of an eye compared with the long history of the Earth. Yet we have changed the face of the planet more than any other species that has ever lived here.

People have great power, and if enough people care about saving the planet, anything is possible. There is an old Chinese proverb that says 'The journey of a thousand miles begins with just a single step.' Do you want to take the first step towards saving your planet? Every little step you take really will make a difference.

# ADDRESSES AND WEBSITES

Find out more about what you can do to help to save the planet by contacting some of the organizations listed below. You can write to them for more information, but please remember to send a large stamped, addressed envelope or a self-addressed label and some stamps.

Friends of the Earth
26-28 Underwood Street
London N1 7JQ
www.foe.co.uk

World Wide Fund for
   Nature
(WWF-UK)
Panda House, Weyside Park
Godalming, Surrey GU7 IXR
www.wwf-uk.org

Greenpeace UK
Canonbury Villas
London N1 2PN
www.greenpeace.org.uk

Royal Society for the
Protection of Birds (RSPB)
The Lodge, Sandy
Bedfordshire SG19 2DL
www.rspb.org.uk/youth

Centre for Alternative
   Technology
Llwyngwern Quarry
Machynlleth
Powys SY20 9AZ
www.cat.org.uk

Sustrans
(sustainable transport)
35 King Street
Bristol BS1 4DZ
www.sustrans.org.uk.

The Soil Association
(organic farming)
86–88 Colston Street
Bristol BS1 5BB
www.soilassociation.org/SA/S
AWeb.nsf/!Open

Tourism Concern
Southlands College
Roehampton Institute
Wimbledon Parkside
London SW19 5NN

Wastewatch (recycling)
Europa House,
13-17 Ironmonger Row
London EC1V 3QN
www.wastewatch.org.uk

The Wildlife Trusts
Freepost, DC526
Lincoln LN5 7BR
www.wildlifetrust.org.uk

Zoocheck
Cherry Tree Cottage
Coldharbour, Dorking
Surrey RH5 6HA
www.bornfree.org.uk/zoocheck

Gaia Foundation
18 Well Walk
London NW3 1LD
Telephone: 020 7435 5000

Aluminium Can Recycling
   Association
I-MEX House
52 Bulcher Street
Birmingham B1 1QU.
Telephone: 0121 633 4656

Tidy Britain Group
The Pier, Wigan
Lancashire WN3 4EX
tidybritain.org.uk

Council for Environmental
   Education
94 London Street
Reading RG1 4SJ
www.cee.org.uk.

The Woodland Trust
Freepost, Grantham
Lincolnshire NG31 6BR
www.woodland-trust.org.uk

Transport 2000
Walkden House
10 Melton Street
London NW1 2EJ
www.llb.labournet.org.uk

Community Recycling
   Network
10-12 Picton Street
Montpelier, Bristol BS6 5QA
www.crn.org.uk

RECOUP
(recycling of used plastic
bottles)
9 Metro Centre
Welbeck Way, Woodston
Peterborough PE2 7WH
www.recoup.org/recoup/